THE MARTIAN

SIDEKICK
TO THE ANDY WEIR NOVEL

by

Allison Clare Theveny

Published by

WeLoveNovels

Disclaimer: This publication is an unofficial Sidekick to *The Martian* and does not contain the novel. It is designed for fiction enthusiasts who are reading the novel, or have just finished. Order a copy of the novel *The Martian* on Amazon.

WeLoveNovels maintains an independent voice in delivering critical analysis and commentary; we are not affiliated with or endorsed by the publisher or author of *The Martian*.

Questions? Ideas? Comments?

Email founders@welovenovels.com.

We are listening!

Table of Contents

INTRODUCTION   7

EXPLORING THE AUTHOR'S FICTIONAL WORLD   11

CHAPTER ANALYSIS & DISCUSSION   13

CHARACTER GUIDE   37

A CLOSER LOOK: MARK WATNEY   39

IMAGINING ALTERNATE ENDINGS   43

THEMES & SYMBOLS YOU MAY HAVE MISSED   47

POSSIBLE STORYLINES FOR A PREQUEL   51

SHOCKS & SURPRISES   55

WOULD YOU HAVE WHAT IT TAKES?   59

IN THE FINAL ANALYSIS . . .   63

IF YOU LOVED THIS NOVEL . . .   67

ABOUT THE AUTHOR OF THIS SIDEKICK   71

Introduction

*L*ost, *Cast Away, Apollo 13...* and *The Martian*? Though the first three stories might differ in medium from the last, they all detail equally riveting and realistic aspects of the struggle to survive. Even if it's lacking the clearly grounded-in-fact polar bears and evil Others of *Lost,* Weir's story ignites that curiosity essential to any tale of survival: can protagonist Mark Watney (could I...) live on and rescue himself (myself...) from Mars?

While I could tell you of the perfection behind Weir's plot of science and surprise, the story behind *The Martian's* publication is convincing on its own *and* allows me to avoid the worst kind of literary faux pas, the spoiler. *The Martian* went from a serial posted in chunks on the web to a ninety-nine cent Amazon e-book that sold oh-so-much better than Mr. Weir ever imagined—and then went on to become a New York Times bestseller, all because fans told Weir he should sell the story that they had grown to love.

What inspired such love? A setting hostile to the protagonist? An author's commitment to scientific accuracy? A struggle for survival and rescue that never once seems to lead to an inevitable conclusion? Though the reasons for the popularity of the book might initially seem as unknowable as its science, any of Weir's readers can tell you he makes the plot and its intrigue as clear as any evening in which we Earthlings can see the little red light of Mars.

How to Get the Most out of This Sidekick

The next two sections of this Sidekick— *Exploring the Author's Fictional World* and *Chapter Analysis & Discussion*—will be most useful <u>while</u> you are reading the novel. You can also refer to the *Character Guide* for a quick "who's who." All subsequent sections—from *A Closer Look* all the way through to *If You Loved This Novel*—are designed to be read <u>after</u> you have finished the novel.

Wherever you are in your reading of the book, with this Sidekick you'll get a chance to dive a little deeper into the author's world, spend some more time with your favorite characters, and check out some of the book's themes and symbolism. There might even be a few Easter eggs along the way.

If you don't have a copy of *The Martian* yet, be sure to pick one up before you get started— after all, here be spoilers! (Don't worry, they're clearly marked.) You can order a copy of the novel on Amazon.

* * *

Disclaimer: I'm not Andy Weir. I don't know Andy Weir. Neither I nor WeLoveNovels has any association with Andy Weir whatsoever. What you've got in your hands is 100% independent and unauthorized opinion, commentary, and analysis.

Exploring the Author's Fictional World

The Martian without Mars would be like *Harry Potter* without Hogwarts or *Game of Thrones* without Westeros. Unlike those tales of adventure, however, Mark's journey occurs through a terrain of fact and not fantasy.

One hundred and forty million miles from Earth and unable to sustain life, the Red Planet functions as both the setting *and* the antagonist in Weir's novel. Unlike most antagonists, however, Mars has no apparent weakness our protagonist can exploit. Instead, Mars reminds us that life is fragile and never lets you forget that everything that lives ultimately dies one way or another, one time or another.

Andy Weir has not traveled to space, but he nonetheless creates a believable Mars, in part because he keeps his imagination firmly grounded in research. As he wrote the novel, he immersed himself in topics such as orbital mechanics, the Martian environment, and botany in order to ensure that the dusty red planet felt real . . . and threatening. Even his writing process invited authenticity, as he published portions of the story to the web in order to receive feedback on both accuracy and style.

Chapter Analysis & Discussion

Chapters 1–3

Mark writes, and writes, and writes about the Hab, the MDV, the MAV, the EVA, and it all seems so plausible. He tells us you can make water from hydrazine, and I, so far from scientifically inclined that I temporarily forgot its so-called "recipe," believe him.

Yet, from the very first page of Andy Weir's *The Martian*, you wonder. The entire premise of the story—man flies to Mars . . . can that even occur, or has Weir firmly rooted us in some fantastical sector of science fiction? While wonder and awe at human ingenuity and survival certainly keep our eyes wide and unblinking, the name of Mark's mission, mentioned, in fact, on that very first page, suggests that we should not place this character and his tale in some distant future. In naming the mission the Ares Program, Weir does not seek to impress us with an allusion to ancient deities. (Mars was the Roman god of war; his Greek counterpart was Ares.) Rather, Weir steals the moniker from a very real NASA venture. Ares I and Ares V were rockets designed as part of the Constellation program, which included a planned human mission to Mars. (The program was cancelled when NASA's funding was cut in late 2010.) So, then, Weir wants us to know *this could happen*, maybe not to those like myself who hope to keep their feet firmly planted on solid, bacteria-rich ground—but to someone.

But to what sort of someone? Perhaps that's the big idea of *The Martian,* a character study of those who "go where no man has gone before" and (possibly) live to tell the tale. I, however, think not. We do not even learn Mark Watney's name until the third chapter, suggesting that this is not a story about a specific man, but rather about mankind, our humanity, and our ability to survive. Look at the end of the first chapter, where Mark relays all the ways he might die: "If the oxygenator breaks down, I'll suffocate. If the water reclaimer breaks down, I'll die of thirst. If the Hab breaches, I'll just kind of explode. If none of those things happen, I'll eventually run out of food and starve to death." While none of these deaths are particularly appealing, none resemble anything fantastic or futuristic. Instead, they center around the essential ingredients of life—air to breathe, water to drink, food to eat, and a body to move in. The fragility of these ingredients and our desperate desire to preserve them at all costs seem to be just what Weir hopes to investigate.

Chapters 4–6

Or is this story about Mark Watney's resilience? Task completion, rather than mental anguish, fills his logs, informing us of the step-by-step process and the potential pitfalls of his plan to grow crops. Never once does Mark, and by extension Weir, focus upon the loneliness or terror of his situation ... except, that is, for ruminations that the Hab is a bomb and he will die a fiery death.

Instead, thoughts about how to create fruitful soil or generate more water preoccupy Mark's mind and perhaps explain his survival. Barring a sudden turn into the sort of story that embraces little green men, I confidently wager that Mark *is* the Martian Weir alludes to in his title. Weir's decision six chapters into the story to provide us with a new point of view stemming from our home turf supports this assumption. When they work out that Mark is alive, the employees of NASA do not believe he can survive, partially because of their own "failure of imagination."

Each of these earthly characters—with their sobbing and screaming, their public facades and presentations that mask their anxieties—inject the very emotional responses into the text that we expected to see, but did not, from Mark himself.

Of course, Weir might simply wish to contrast the monotony, the emptiness, the nothingness that is Mars with the panic of Earth. The non-Mark chapter depicts a tizzy of interaction, planning, and emotion as the characters worry about how to communicate their shocking news to the public. At the same time, Mark *needs* to communicate but cannot, and Mark very well might need an emotional outlet, but he has none. In fact, we might consider that the only way we hear Mark's voice at all is through his logs: a conscious recording and presentation of his thoughts, not unlike Annie's PR control and Teddy's speeches. Perhaps we see emotion on Earth because we experience events as they develop, rather than after someone has processed and distilled them for wider consumption.

Even so, I felt Weir's joke coming. While the non-astronauts, safe on Earth, wring their hands and worry about Mark Watney's psychology, our Martian takes a little mental break with an episode of TV, bucking up for a tomorrow in which everything *but* a failure of imagination will stop him.

Chapters 7–9

If you found yourself puzzling over Mark's clever name for his missions, you're in good company. Of course, I absolutely knew (wink, wink) that Sirius refers to the ball of gas known as the Dog Star, named because of its prominence in a constellation known as Canis Major, which both the Greeks and the Romans thought represented various dogs of their mythology. But I didn't understand what Mark's mission had to do with dogs or Harry Potter's godfather . . . until I realized that Mark wants to test the limits of the *rover*, a machine that shares its name with the most cliché of pets. This kind of humor is fast proving to be one of Mark's most defining character traits—well, that and his single-minded determination to survive. Certainly, Dr. Irene Shields has noticed it, and she implies that Mark represented the glue of the Ares 3 mission. For all his good-natured optimism, however, Mark does not seem to share this interpretation of his place on the team. From the beginning, he introduced himself as the lowest-ranking member of the crew, and even now, as he marvels at being the first to do so many things on Mars, he says, "I wasn't expected to be first at anything." But we, the psychologist, and Weir know that Mark is special, right?

Yet, as I return to a particular passage, I cannot help but allow a tiny shred of doubt regarding our protagonist to creep into my head. Venkat remarks that "everyone's pulling to save Mark. . . . [This] is going to cost tens of millions, maybe hundreds of millions of dollars." It's easy to cheer Mark on, to hope for his survival. But this sentence offers us a momentary chance to play devil's advocate. Are those tens of millions best spent rescuing a single man, or would they be better used to save hundreds—perhaps thousands—of lives on Earth?

The dilemma is firmly rooted in history. These days, most people associate the moon landing with wonder, but at the time, many (especially the young and flower-power inclined) believed that the government should have spent the money on the abundant problems at home before looking up into the sky. Now, in Weir's story, we have a single, precious human life in jeopardy, and without hesitation, NASA has devoted tens of millions of dollars to a project that may fail to save it. A thorny philosophical issue for us to ponder as we continue to read and, if you're interested, to explore even further by watching the episode of *Mad Men* featuring the moon landing ("Waterloo").

Chapters 10–12

Finally, more astronauts—and the foil to Mark's easy-going mentality! Commander Lewis emerges as self-sacrificing and self-blaming, possessing basically every character flaw (or strength) we have come to associate with *the hero*. In fact, if your book cover looks anything like mine, you'll have noted the big bold letters informing us that *The Martian* will soon become "a major motion picture starring Matt Damon," and perhaps you've thought that a character like Lewis was long overdue in a novel sure to become a future summer blockbuster.

But Weir keeps insisting that Mark is different. After all, it takes him 115 pages to acknowledge that he has "spent three months as the loneliest man in history." His most basic instinct—to take risks in order to survive—directly contrasts with Lewis's impulse to risk her life in order to save others. Which begs the following question:

Though he does not ask this question directly, Weir does infuse *The Martian* with thought-provoking ethical dilemmas. There's little chance of success, yet Mark and NASA put all their resources toward Mark's survival. Five people are devastated by the loss of a friend and crew member, but Venkat refuses to relay information that might result in a potentially deadly distraction for those five people . . . or is it for five trained and valuable astronauts flying a very expensive ship?

Such questions, however, cannot fully distract from the structural genius of these chapters. Mark and Earth struggle to communicate using a slow and cumbersome process, and then, WHAM!, Mark and the crew banter back and forth rapid-fire in a sudden, unexpected flashback sequence. Rather than allowing Mark or the text to dwell on the loneliness of the protagonist's situation, Weir communicates that feeling to us via a meticulous—at times almost boring—summary of Mark's day. It presents a striking contrast with the vibrancy, vitality, commotion, and chaos that the other characters experience. Weir shows us that life, though hard, occurs with others. When alone, you become an alien. A Martian. A Mark Watney.

Chapters 13–15

That pesky problem of money arises again. What's the cost of survival? That's the question Weir asks in an uncharacteristically direct manner over these next few chapters. Consider the following:

> "*The overtime pay was spectacular, and the funding was limitless.*"

> "The attention is good. It'll help us get emergency funding from Congress."

> "Everyone, I understand your positions. We have procedures. Skipping those procedures means risk. Risk means trouble for your department. But now isn't the time to cover our asses. We have to take risks or Mark Watney dies."

> "Is there any spending limit to this rescue operation? . . . Some people are beginning to ask how much is too much."

"It's not about the bottom line. . . . It's about a human life in immediate danger. But if you want to look at it financially, consider the value of Mark Watney's extended mission. His prolonged mission and fight for survival are giving us more knowledge about Mars than the rest of the Ares program combined."

Of all the themes bubbling beneath the surface of *The Martian*, Weir seems intent upon directly articulating the tension between the value of life and the value of the dollar. Elsewhere in these same chapters, Weir hints more subtly at this tension. For example, Mark considers the email from his mother infinitely precious, but he kids that the mothers and sisters of NASA's investigation committee members are prostitutes. The two very different references to mothers in this series of chapters make an interesting point: Mark values his mother because she is *his* mother. By connecting mothers to prostitutes, Weir asks, Do certain lives have more value than others? NASA tries to have it both ways, stating that Mark's life is about more than the bottom line while simultaneously noting that Mark's presence on Mars has enormous value. In a way, Weir's forcing us to consider that dichotomy by showing us a corporation that has seemingly embraced both answers. It's almost as if Weir has wondered if we the reader can rest easy knowing that both philosophies have justified the decision to save Mark. What if Mark's time on Mars had no value to NASA?

Chapters 16–18

Diplomacy, mutiny, and a reference to *The Lord of the Rings?* Did I just turn the page and begin another novel?

Most likely, Weir has suddenly injected these intense plot lines into our narrative for a very specific reason. Power plays and backstabbing, whether made from across the world or across the universe, involve human relationships. Even the reference to Elrond highlights human (and elven, and dwarvish, and wizardly) connections to one another and to a larger purpose. But just as the narrative becomes, in this sense, its most social, Weir once again cuts Mark off from the rest of the world. The sudden silence he (and we) experience is deafening.

This makes Weir's decision to show us a more emotional side of Mark's character all the more significant. By interspersing Mark's personal letters to his crewmates throughout these chapters, and by reminding us that someone will have to speak with Mark's parents should he die, Weir begs us not to forget Mark's humanity, no matter how inhuman his ability to defy the odds.

Yet the book does not convey hopelessness or generate needless drama. Just a couple paragraphs before Mark admits, "I fucked up," he realized that some of the bacteria, despite exposure to the cold, had survived. The juxtaposition is striking, especially as it marks the shift between social connection and utter isolation. "It only takes one survivor" may apply to Mark every bit as much as it does to the bacteria—and the sentiment provides a hint of hope that isolation need not be the end for our hero.

Chapters 19–21

Weir lets Mark ask the most important question of the novel: "What would an Apollo astronaut do?" The answer Mark provides hints at what Weir might wish to accomplish with *The Martian*. The James Bond astronaut he describes, the one who would "drink three whiskey sours [and then] drive his Corvette to the launchpad" exudes coolness. Though he describes himself as a nerd, more of a Q than a 007, Mark often appears calm and unflappable in the face of danger—even if he doesn't see the Bond parallels, we sure do.

Does this simply point to the sort of character that Weir wished to create? Other portions of these chapters suggest otherwise. Contextualize this Apollo astronaut, this James Bond figure with statements like, "It's amazing how much red tape gets cut when everyone's rooting for one man to survive," and with plot details like the Chinese desire for a strong space legacy of their own. It seems likely that Weir wants to attribute a coolness to Mark because he believes space itself is cool, and, more importantly, because he does not want any more "net loss[es] for mankind's knowledge" to occur.

Keep in mind that the name "Ares" references those rocket designs that got scrapped when NASA's funding was cut in 2010. We must remember this context and ask why: Why did Weir choose Ares as the name of his imaginary program? Why did he place the story in the near future? Why did he so accurately present the science of his work? Perhaps the answer is that Weir is not just a storyteller, but also an advocate—one who thinks space exploration has an important role to play in our world.

Chapters 22–24

ALERT: SPOILERS BELOW

In Chapter 24, Mark starts daydreaming about a beautiful Queen of Mars and says, "It's been a long time since I've seen a woman"—and can't go on with his narrative without reiterating the sentiment. It's worth considering the way in which Weir has (and has not) used gender throughout *The Martian*. On the one hand, we have Mindy, Annie, Lewis, and Johanssen, all good at their jobs and all distinguishable as characters. On the other hand, we have casual comments about mothers as prostitutes and a "that's what she said" joke, all coming from our male protagonist. Beyond these jokes that do, admittedly, complicate the otherwise strong writing of women as believable human beings in their own right, *The Martian* seems neither unconscious of gender nor preoccupied with its implications. Perhaps Weir includes these jokes to highlight gender's lack of importance on a lifeless planet. The jokes Mark makes, and the norms and stereotypes they convey, are earthly; they matter to the story only when heard by those from home.

This penultimate section of chapters throws dust in our eyes literally and figuratively. Though Mark's survival still doesn't seem like a foregone conclusion, though he has no way of knowing of the dusty danger that lurks above and in front of him, and though he cannot prevent his rover from flipping, Weir uses little hints to suggest that all might work out, miraculously, as intended. As Mark attempts to outsmart the storm, he thinks about the grandchildren to whom he will tell his no-need-to-be-exaggerated tale of woe. Similarly, Martinez tells Lewis, and by extension us, to "have faith in Mark Watney."

Weir even adopts an anonymous third person perspective. We saw this once before, when the air lock breach was imminent. Then he used the shift in narrative style to build drama. This time, he romanticizes and dramatizes the journey of "the traveler," further solidifying Mark Watney as the hero of this adventure and the (I still dare to hope) survivor of this tale.

These glimmers of hope may have nothing to do with the ultimate outcome of Watney's struggles. Instead, they might simply represent another of Weir's lessons on what it takes for humans to survive. If you do not have hope, Weir argues, if you do not believe that the next moment will get better, then you will not make it even one moment after some antenna flies out of nowhere, knocks you down, and impales you.

Chapters 25–26

Loud and clear, Weir's screaming: you are Mark, and Mark is you. Mark has not showered; he smells like any man would. Mark has a daily routine (complete with TV) to get through his days, a routine that ultimately works. When he breaks two ribs, Mark "scream[s] like a little girl." Over the course of the novel, Mark has been a farmer, a trucker, a construction worker—all professions representative of the ordinary, blue-collar everyman. He's also been a scientist, an engineer, an explorer ... and ultimately, of course, a survivor.

The act of saving Mark comes straight out of an action movie, but the motivations are ordinary. These characters think, *I will survive, and we will save our friend*, and then they use their incredible brains to do it. Even though the astronauts make bombs and throw themselves out into space to do the actual saving, Weir nonetheless ties these actions to something universal: "This is so fundamentally human that it's found in every culture without exception. Yes, there are assholes who just don't care, but they're massively outnumbered by the people who do." The urge to help, he argues, is innate.

Weir's thematic statement finally becomes evident on the very last page of his book. He seems confident that we will hear this story and agree with its sentiment, but I wonder if countries, big organizations with big money, and the average human being really would risk it all for one man. Perhaps that makes me more like the Teddys of the world than the Martins, and perhaps that's part of Weir's point. While I wish it provided me with soliloquy after soliloquy on humanity's urge to help, *The Martian* shows us that survival has little to do with, as Mark puts it, "waxing philosophical"—and everything to do with action after risky action. In other words, *The Martian* shows, rather than tells, us that when crises erupt, most people do not stop to ponder; they just act.

Character Guide

*A*lex Vogel – Ares 3 German mad scientist.

Annie Montrose – Director of media relations and publicity guru for NASA.

Beth Johanssen – Youngest Ares 3 crew member; a software engineer and radar and sysop tech.

Chris Beck – Ares 3 doctor, biologist, and EVA specialist.

Mark Watney – Our Martian, our hero.

Melissa Lewis – Ares 3 commander, geologist, and lover of disco; U.S. Navy.

Mindy Park – NASA orbital engineer responsible for monitoring the Martian satellites.

Mitch Henderson – Ares 3 flight director.

Rick Martinez – Ares 3 pilot; major in the U.S. Air Force.

Teddy Sanders – Administrator of NASA.

Venkat Kapoor – Director of Mars operations for NASA.

A Closer Look: Mark Watney

Mark Watney: Man or Martian?

As the NASA psychologist states, resourcefulness and humor define Mark Watney, the lone man, the Martian struggling to survive. But what of Mark Watney, the Earthling?

Though we know that Mark's attitude directly and positively impacts the morale of his crew, we know little about how he relates, in a deep and real sense, to friends and family. Weir does sprinkle little clues throughout the novel: Mark clings to the letter from his mom, reading it over and over again; he writes Martinez, informing him that he feels he can ask him to speak with his parents if he does not survive; and he recognizes that Beck has fallen in love with Johanssen. But though we know Mark has emotion and is emotionally intelligent, we do not ultimately receive much insight into Mark's daily emotional state. Instead, we as readers must decide if he simply withholds his fear and anger from his logs or if he, as a person, forever hides that fear and anger from all, even from himself.

What little we do know raises more questions. While we do not know Mark's age, the fact that he speaks only of his mother, father, and extended family throughout the novel implies he is unmarried. If you watch the trailer for *The Martian* movie, you might notice that the director seems to have decided to give Watney a wife and child, presumably to pack a greater emotional punch. I would argue, however, that Mark becomes even more accessible as a character if he has no family of his own. Perhaps we are not meant to know Mark as a person. Perhaps Weir intended Mark as a blank slate, a way to enter the story and wonder if we, too, could survive.

Though it's intended as a joke, the moment in which Mark remarks that he wishes to act like the Apollo crew that drinks whiskey sours and drives corvettes ultimately indicates that Mark has a vision of who he—astronaut, survivor, and man—should be. We don't get to see much beyond this vision, because Mark does not show us or the people on Earth his vulnerability—and perhaps he cannot, if he is to stay strong and survive.

The Martian
A Sidekick to the Andy Weir Novel

Imagining Alternate Endings

Left Behind Lewis

The crew takes off, leaving not one, but two Martians behind. Martinez tries but ultimately fails to reverse the MAV's tilt, so their commander and the presumed dead Mark Watney together "colonize" Mars. Lewis sharpens and streamlines many of Mark's plans and causes one or two fires of her own, but the plot details of their survival do not really change. At least, not at first.

Successful communication with Mother Earth plants a black seed deep in the commander's mind. The proposed rescue plans waste resources that could save lives on Earth, decimate funding for future NASA missions, and might even put the crew she sacrificed everything for back in jeopardy. She could kill herself to prevent disaster. After all, it's just one (okay, two) lives.

Mark Watney must survive Mars, but now he has a new adversary: Commander Lewis and her arguments for suicide that might just turn murderous. Would Lewis convince (or force) Mark before Hermes reroutes? Would Mark kill Lewis in self defense? Would both survive, with Lewis's fragile psyche and dangerous plans exposed, or would Mark stay silent, allowing the captain who would sacrifice anything for the greater good to become a hero . . . or is she one regardless? That imagining I leave to you.

Houston, We Have a Problem

The Rich Purnell Maneuver fails, and immediately Lewis, Martinez, and Vogel take their pills. Beck, however, hesitates. As Beck attempts to say goodbye to the woman with whom he has fallen in love, Johanssen rips the pill from his hand, saying she can't let him go yet, that they can stretch the food supply at least for a little while longer. Beck argues, but half-heartedly—for now that the moment has passed, his friends bodies lie on the deck, and death seems much more frightening.

They continue on and time passes, with Earth becoming smaller and the little pill seemingly becoming bigger. Johanssen reasons that where she alone would fail, the two of them together can still save Mark. They can worry about food later, and that pill isn't going anywhere.

But even together, Beck and Johanssen fail. Beck reaches Mark, but they miss the ship and drift in space until they die. A despondent Johanssen wishes to die, too, and eventually realizes that her food supply will not last. She takes the pill, and the unmanned ship crashes.

Millions of dollars to save one astronaut, and six die in the attempt.

Themes & Symbols You May Have Missed

Fire

Mark jokes that almost all of his inventive stay-alive schemes have used fire. While the oldest of parental admonishments involving playing with matches reminds us to laugh at Mark's remark, it also draws our attention to a very specific literary flourish.

To create the water he needs to grow food, Mark needs fire—yet that same process almost results in a fiery explosion that will kill him. More than once, Mark needs a source of heat and must turn to the oh-so-dangerous, bury-it-four-miles-away ball of radiation. Finally, each time Mark makes fire, he must circumvent NASA's attempt to strip their life-giving systems and machines of flammable materials.

This dichotomy between fire as life-giving and fire as death-inducing isn't confined to *The Martian*—we find it all throughout art and life; it's a repeating motif. Frankenstein's monster discovers his own humanity in a variety of ways, but his most striking moment of discovery occurs as he makes fire, experiencing its warmth and perceiving its danger. Prometheus stole fire from the gods, giving humans great power but resulting in his personal ruin. *The Martian* continues in this tradition, telling us that to live is to risk life. Fire, so essential to humanity that the cavemen's discovery of the flame has become a cliché, perfectly encapsulates such a message.

Disco

Yes, disco. While disco is a running joke throughout the book, the comedy has hidden depths. We don't need to count the number of snide remarks that Mark makes about disco to know that he does so *a lot*. Consequently, we need to ask, of all the bad music out there, why did Weir select disco? Did he simply want to work in that "Stayin' Alive" joke? Maybe.

Or maybe not. The (near) universal mockery of disco abounds here on Earth, yet on Mars, that means nothing. There are no humans with whom Mark can share any thought or opinion, no matter how universal or controversial. Disco, and his hatred of it, keeps Mark tied to Earth and its culture. Without it, his jokes—and his ability to cope—might disappear. Lewis's conversation with her husband strengthens this idea; they bond over disco while miles and miles apart from one another. Similarly, Mark's complaints about disco endear him to the NASA officials working to save him. It reminds them of who he is as a person, reminds the reader of his humanity, and perhaps even reminds Mark, the Martian, of his essential humanness.

Possible Storylines for a Prequel

The Martian depicts the survival of the seventeenth person to set foot on Mars, but what of those that came before? Not Lewis and the like, but the very first humans to explore the Red Planet.

Were Ares 1 and 2 a success? Certainly, they were successful enough that the Ares program continued, but what problems did those astronauts encounter and what secrets did they keep among themselves? Weir wanted *The Martian* to stick to the science part of science fiction, but what if a prequel contained a bit more of the fiction?

Imagine this: Ares 1 arrives on Mars and ordinary human beings step onto its dust for the first time in history. Whether from radiation from improper use of the RTG or from contact with some foreign Martian substance, a reaction ensues that changes the astronauts' internal makeup, a la Spiderman! With new and strange abilities, they return to Earth, where they must conceal their true natures from all.

Ares 2 lands not long after, and what do they find but the elusive extraterrestrial! Neither as friendly as ET nor as frightening as those things in *Alien*, this particular extraterrestrial hails from another corner of the universe, has known all about Earthlings for a while now, and just *hates* that we set foot on Mars first. Our astronauts learn much about our universe before departing, but little that those back on Earth would believe or understand, and so a secret it must remain.

Of course, Weir could also write a prequel about the previous Ares missions, documenting their extraordinary (but fantasy-stripped) struggles and successes. If it's anything like *The Martian*, that would definitely entertain, no ETs or Spiderman required.

Shocks & Surprises

Wait, What, Cannibalism?!

Is it just me, or can we discuss this novel without mentioning the moment when Johanssen reassures her Dad with the classic, "Don't worry, I'll survive by eating everyone else!"? The inclusion of this detail shocked me–so much that I gasped aloud in a café and desperately wanted to make friends with the dude who was staring at me just to tell him what had occurred.

Once my eyes could return to the page, a number of explanations for why Weir included this moment began to flood my brain, including (1) cannibalism represents a realistic option considered by astronauts and sanctioned by their training, (2) Weir wanted to stress the full extent of an astronaut's psychological and physical commitment to her mission, or (3) all survival stories, whether or not they involve a desert island, must allude to the classic cannibal trope. Personally, I am partial to third, did not really want to know if there was any truth to the first, and figured the second played at least some part in Weir's decision. What say you?

Weir Published This Book How?!

Can just anybody drop a book out into the void that is the Internet and one day wake up to find that the New York Times bestseller list has a new number twelve?

When I first learned that Weir had placed a version of *The Martian* on his personal website and only later attracted the attention of publishers, I thought I had heard a story akin to

The Little Engine That Could, a David-like author overcoming the Goliath-sized publishing industry.

But I was wrong. As surprising as I found *The Martian*'s origin story, I should have recognized its achievement as standard for the twenty-first century. One of the most popular and hotly debated books of late, *Fifty Shades of Grey*, started as *Twilight* fan fiction. Publishers guessed it would bring in the big bucks with a bit of polish and a few name changes. Though admittedly having more in common genre-wise with *Fifty Shades* than *The Martian*, *After* is another fanfic that will turn into a big-screen flick; it centers around a young woman who falls in love with a member of One Direction. Of course, online . . . literature (whether the term applies is a debate for another time!) shares this new pathway to success with many other art forms. Take film, for example: Zach Braff wanted to make another *Garden State*-esque film, so he turned to Kickstarter. Even YouTube singers and entertainers can secure record deals—or, at least, widespread attention—if they manage to attract viewer after viewer after viewer.

These days, popular will governs much more than the book or movie of the year. Uber, the taxi-like service you can call on your iPhone, and Airbnb, the online service that allows you to pay someone to use their home while you're traveling abroad, are just two examples of how technology has started giving *us* the choices and the control. Industry pushback does occur (Uber and Airbnb, for example, are getting sued for failing to comply with regulations, taxes, and the like), and might find a way to regain the little control they've lost. Even so, while the publishing industry might find a formula to predict which online books will automatically translate into offline successes, we can still admire how very different books, like *The Martian* and *Fifty Shades of Grey*, found their audiences before "the man" made his mark on their pages.

Would You Have What It Takes?

Mark seems like the sort of dude with whom you could grab a beer. He's funny, not too high strung, and just plain likeable. He seems a bit like the everyman.

But does the everyman become an astronaut? I think not. Throughout the entirety of the novel, I wondered at Mark's intelligence and, even more, at his emotional stability. At one point in the novel, after a NASA psychologist states that all the astronauts selected are smart and stable, I immediately realized what most people probably already knew: not just anybody can sign up for the next shuttle to space.

So what does it take to become an astronaut? Predictably, a lot of unsurprising requirements like a bachelor's or (even better) an advanced degree in something science-y, like engineering, biology, physics, or mathematics; a thousand hours of pilot-in-command time in a jet aircraft (or three years of "professional related experience"); and the ability to meet certain physical standards related to vision, blood pressure, and height.

Then, there are the more elusive, psychological standards. NASA officials have stated that they look to weed out the wrong stuff almost as much as they search for the right stuff. In other words, they want men and women of resilience who possess no mental illnesses and no personality issues that would cause problems on long missions. How exactly does NASA determine this resilience or smoke out potential instability? Interviews, of course, and perhaps stress tests, but there's no concrete, publicly available info as to the specifics or the full procedure behind this selection process.

Of course, it may also depend on the mission. On June 16, 2015, six NASA scientists emerged from a dome set up atop a dormant volcano in Hawaii after spending eight months in the enclosure. The purpose of the extended stay: to simulate a Mars-like environment and determine the effects of isolation on team dynamics. That's something like 248 days in a Hab-like structure. Could you make it?

The Martian
A Sidekick to the Andy Weir Novel

In the Final Analysis . . .

The Martian is an unusual novel, and not just because of its extraordinary path to publication. In some ways, the first half of the novel reads like a fact-driven, how-to guide . . . if your standard how-to-guide covers surviving on Mars, that is. Then Mark connects with Earth, and the story shifts, becoming faster in pace and more character driven.

At first, I wondered at this structure. I felt that Mark's logs, particularly near the beginning and the end of the book, were too concerned with the details of science for my particular preference for soft science fiction. I asked again and again, If our protagonist is alone on a planet that cannot sustain human life, where is the angst?

Over time, however, I grew to love Weir's refusal to dig deep via flowery language or internal monologues. Like a film or a piece of art, *The Martian* conveys theme and character without telling its audience exactly what to make of the moments it depicts. Mark's meticulous detailing of his day-to-day life on the planet imparts boredom, loneliness, and an understanding of the complexity of sustaining life without ever actually articulating these ideas. Such detail-driven writing resembles classic adventure stories like *The Odyssey*, in which the events that occur seem to matter so much more than the authorial point behind those events. This style allows us to become like Mark, alone in the universe, trying to figure out how we feel and what we think about what is happening to us.

Furthermore, this lack of heavy-handedness renders the moments in which Weir does become abstract all the more significant. For example, the arguments between Mitch and Teddy about risk articulate a hidden meaning behind Mark's struggle—that the risk we strive to avoid often defines life. The conclusion of *The Martian* perfectly summarizes this subtle literary intent. In remarking that humans, across all cultures and groups, have an impulse to help other humans, Mark challenges us to consider this abstract truth against the backdrop of what we just read, against what we ourselves have experienced. Yet while such themes add depth and interest to the story, it remains at its core just that: a story—a compelling narrative in which all messages Weir might wish to impart take a back seat to the day-to-day process of Mark's survival.

The Martian
A Sidekick to the Andy Weir Novel

If You Loved This Novel . . .

The Egg, by Andy Weir

Same author, completely different story, both structurally and thematically. Where *The Martian* deals in survival, *The Egg* deals in death: the death of the protagonist, to be precise. The first of Weir's published works, and the shortest, *The Egg* manages to contain more overt philosophy in its few words than *The Martian* does in its entirety.

The Martian Chronicles, by Ray Bradbury

Though they zero in on the same setting, Bradbury and Weir differ widely in how they depict human contact with Mars. *The Martian Chronicles* is a series of short stories detailing the human colonization of Mars, and Bradbury's format, tropes, and themes offer a good point of contrast to *The Martian*. Plus, if you were disappointed by the lack of little green men in *The Martian*, Bradbury just might give you what you're looking for.

The Mars Trilogy, by Kim Stanley Robinson

Robinson's trilogy (*Red Mars*, *Green Mars*, and *Blue Mars*) tackles the topic of otherworldly survival on a vast scale, chronicling the settlement of Mars through a variety of characters' perspectives over the course of several centuries. Despite this difference, Robinson, like Weir, strove to create a novel of hard science fiction in which accuracy and detail matter as much as plot, character, and the like.

So, What'd You Think?

Thanks for investing in this *Sidekick*. Now that you've read it, let us hear from you!

In just a sentence or two, please email founders@welovenovels.com your answer to one simple question:

What was your favorite (or least favorite) thing about this Sidekick?

We want to know what you think, so we can bring you more of what you love most, and fix what you don't like.

And if you would like a free copy of Katherine Miller's top-rated *Sidekick* to *Leaving Time*, Jodi Picoult's latest bestseller, we'd like to send it to you (a $4.99 value). All you have to do is add the words "**Yes, I Want My Bonus Sidekick**" to the email subject line, and you'll get instant access.

The Martian
A Sidekick to the Andy Weir Novel

About the Author of This Sidekick

Allison grew up in Seattle, WA, and Philadelphia, PA, and studied English and film and media studies at Georgetown University. Upon confronting the post-collegiate "real world," she knew she needed to find a way to continue analyzing (and enjoying!) novels on a regular basis. Consequently, she is excited to share her thoughts with you.

Other Sidekicks from WeLoveNovels

Sidekick to The Nightingale

Sidekick to Wayward

Sidekick to Seveneves

Sidekick to Departure

Sidekick to Orphan Train

Sidekick to Papertowns

Sidekick to Gathering Prey

Sidekick to Pines

Sidekick to Memory Man

Sidekick to The Shadows

Sidekick to The Husband's Secret

Sidekick to A Spool of Blue Thread

Sidekick to The DUFF

Sidekick to Insurgent

Sidekick to Redeployment

Sidekick to The Girl on the Train

Sidekick to Still Alice

Sidekick to Captivated by You

Sidekick to Catching Fire

Sidekick to Mockingjay

Sidekick to Deadline

Sidekick to Big Little Lies

Sidekick to Gone Girl

We are so grateful to all who have taken a moment to leave a quick review of one of our Sidekicks on Amazon. Your thoughtfulness means a lot and helps us, and the rest of the world, know how we are doing and how we can improve. :)

Questions? Ideas? Comments?

Email **founders@welovenovels.com**.

We are listening!